AC Electrics

Classes 91, 92 and DVTs

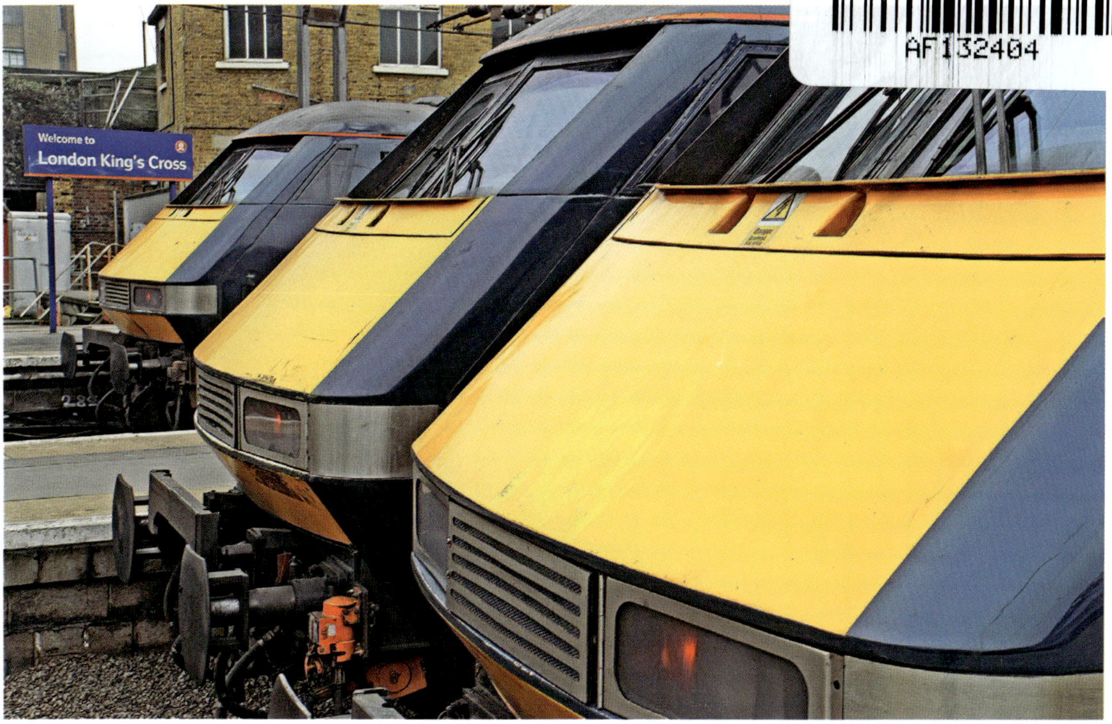

MARK V. PIKE

KEY Books

BRITAIN'S RAILWAYS SERIES, VOLUME 62

Front cover image: Apart from the unidentified HST power car to the right, we have an almost full house of 91s here at King's Cross. From right to left are 91010 + 91007 *Ian Allan* + 91011 + 91023. All of these examples, apart from 91023, are still operational at the end of 2023 but carrying the numbers 91110/107/111. Incidentally, 91023 was involved in two serious accidents in the early 2000s and, as a consequence, when it was refurbished, instead of taking up the number 91123, it was instead given the number 91132 (there was never a 91032). 22 July 1992.

Title page image: This is 91122+91111+91125 all in National Express/GNER livery at King's Cross, having arrived with trains from the north. 20 February 2008.

Back cover image: 92009 is heading north at South Kenton on the West Coast Main Line with 6B41, the 11.43 Wembley to Daventry. Upon repainting, the loco lost its Elgar nameplates, but a short while later it received the cast nameplates *Marco Polo*. It has, however, been out of service for some time now. 27 May 2011.

Published by Key Books
An imprint of Key Publishing Ltd
PO Box 100
Stamford
Lincs PE9 1XQ

www.keypublishing.com

The right of Mark V. Pike to be identified as the author of this book has been asserted in accordance with the Copyright, Designs and Patents Act 1988 Sections 77 and 78.

Copyright © Mark V. Pike, 2024

ISBN 978 1 80282 826 9

All rights reserved. Reproduction in whole or in part in any form whatsoever or by any means is strictly prohibited without the prior permission of the Publisher.

Typeset by SJmagic DESIGN SERVICES, India.

Contents

Chapter 1 Class 91 4

Chapter 2 Class 92 49

Chapter 3 Mk4 Driving Van Trailers (DVTs) 73

Right: 92015+92009 are seen here stabled at Stafford. 23 January 2009.

Below: This is 82200 with its Remembrance Day wrap awaiting departure from Cardiff Central with 1W93, the 11.25 Cardiff Central to Holyhead service. Note the all-over black livery adopted by TfW for these trains. 14 March 2023.

Chapter 1

Class 91

A total of 31 Class 91s were built for British Rail (BR) between 1988–91 for use on the East Coast Main Line (ECML) after its electrification between London King's Cross and Edinburgh. In this way, a few high-speed train (HST) sets already working these services were replaced, but many remained working alongside these new trains for a number of years, especially on services to places such as Aberdeen and Inverness that are not electrified. The new trains became known as Inter-City 225s due to being capable of 140mph (225kmh) running. However, the trains only ever run at a maximum of 125mph in passenger service, due to numerous items of infrastructure not having been upgraded.

In later years, and as a consequence of privatisation, the trains underwent many changes of operator and colour schemes; firstly GNER in 1996, then National Express East Coast from December 2007, then East Coast from November 2009, then Virgin Trains from 2015 and finally, London North Eastern Railway from 2018. Eventually, both the HSTs and the 91s and Mk4 stock were replaced from 2020 by Class 800/801 'Azuma' units on the King's Cross to Edinburgh/Aberdeen/Inverness services. LNER does, however, retain a few locos and stock to work London to Leeds and York services but, with the announcement by LNER in late 2023 that new tri-mode trains are on the way, the clock is ticking for these too.

Starting off with the Class 91s, we are at King's Cross station for a series of 20 shots of the original Class 91/0 configuration as built. Just under two years old, this is pioneer 91001 *Swallow*, awaiting departure with a northbound express. This loco received its flimsy 'tin'-style nameplates in September 1989. It is still in service at the time of writing (December 2023) but now carries hybrid LNER/Inter-City livery similar to that portrayed here. 28 February 1990.

Of all the liveries the class has carried down the years, this original is my personal favourite. 91010 is captured about to depart with what I believe is a service to Leeds. This loco was also still in service at the end of 2023 and has become something of a celebrity, firstly due to it being the record holder of the fastest speed attained by an electric loco in the UK of 162mph, but secondly that it also now carries a rather superb one-off livery. See later in this section. 27 July 1992.

Here we see 91008 *Thomas Cook* departing. There had to be a first loco of the class to be withdrawn and such was the fate of this one in July 2019 (as 91108); it was reduced to scrap during 2022. 26 May 1993.

91026 is seen next to the previous generation of high-speed East Coast Main Line (ECML) traction in the form of high speed train (HST) power car 43109. From the early 1990s right through to the late 2010s, these trains worked side by side on Anglo-Scottish expresses. The Class 91 (as 91126) was withdrawn in 2020 and scrapped in 2021, while the power car is currently in storage (as 43309) at Ely Papworth Sidings with a somewhat bleak future ahead of it. 26 May 1993.

The last of the class, and the final loco to be built at Crewe Works, is 91031 *Sir Henry Royce*. This loco can be seen wearing a small plaque behind the cab, which acknowledges that it was, for a while, the fastest loco in the UK, attaining a speed of 154mph in 1995. This was later superseded by 91110 as mentioned earlier. The plaque was subsequently removed from 91031, but an updated version was later affixed to 91110. This loco (as 91131) is now preserved as a static exhibit at Bo'ness (West Lothian), Museum of Scottish Railways. 22 July 1992.

King's Cross has always been a good place to see groups of locos over the years, especially 91s, as they were almost always attached to the northern end of trains. Here are 91025+91009+91013. Of these locos, 90025 (as 91125) was scrapped during early 2023 and 91009 (as 91109) is still in service in late 2023, but 91013 (as 91113) was reduced to scrap in 2021. 26 May 1993.

Right: On the same day as the previous shot, we see 91016+91013. 91016 (as 91116) was scrapped in early 2023. 26 May 1993.

Below: A few years later, 91027+91007 *Ian Allan* are at the same spot. Still operational at the end of 2023, both 91027 (as 91127 *Neville Hill*) and 91007 (as 91107 *Skyfall*) are now in a hybrid LNER/Inter-City livery which is very close to the original Inter-City colours seen here. 29 August 1996.

91022 *Robert Adley* and 91003 *The Scotsman* are alongside 43110 *Darlington*. Notice the 'real' cast nameplates affixed to the 91, which are sadly lacking from many of the railway vehicles we see named today. 91022 (as 91122) was scrapped in late 2022, while 91003 (as 91103) was the second loco of the class to be withdrawn and was scrapped in 2022. 29 July 1994.

Apart from the unidentified HST power car to the right, we have an almost full house of 91s here! From right to left are 91010 + 91007 *Ian Allan* + 91011 + 91023. All of these examples, apart from 91023, are still operational at the end of 2023. Incidentally, 91023 was seen as something of a jinx as it was involved in two serious accidents in the early 2000s and, as a consequence, when it was refurbished, instead of taking up the number 91123, it was instead given the number 91132 (there was never a 91032). 22 July 1992.

Back to single locos now for the rest of this opening sequence at the capital. This is 91028 (formerly *Guide Dog*) that had just been de-named a few days earlier and was to be repainted into GNER dark blue livery the following month. The loco was scrapped (as 91128) during 2022. 3 April 1997.

It could be surprisingly tricky to photograph the 'blunt' end of these locos, but I was lucky enough to capture 91024 *Reverend W Awdry* reversing down on to its train during a loco change. This view shows to good effect the unusual, but very sensible design that had been developed from the HST power cars, which had just one driving end. This loco is still operational (as 91124) and still doing the same work it was designed for up and down the ECML. 3 April 1997.

The second loco to be withdrawn from service, 91003 *The Scotsman* looks smart here at the head of its train and carrying those impressive nameplates. The loco was scrapped (as 91103) in 2022. 29 July 1994.

One of the station staff has just affixed the temporary 'not to be moved' board to 91014 *Northern Electric* soon after it has come to a stand with a train from the north. This is another of the lucky examples (as 91114) that is still in front-line service at the end of 2023, now carrying hybrid LNER/Inter-City livery and the cast name *Durham Cathedral*. 25 August 1993.

Displaying another superb cast nameplate, 91009 *Saint Nicholas* awaits departure. This loco is also still in traffic (as 91109) but now carries the name *Sir Bobby Robson*. 26 May 1993.

The driver looks ready to go, as a crew member (driver's assistant/second man?) looks back to see what the delay is to 91029 *Queen Elizabeth II* just prior to departure. This was one of the few examples to carry just one name (although of different designs) throughout its life. It was, however, scrapped (as 91129) during 2021. 22 July 1992.

Another huge cast plate is on show here as 91017 *Commonwealth Institute* has just arrived at the station from the north. As of the end of 2023, this loco (as 91117) is in Europhoenix livery; originally intended for export, it is currently located at Barrow Hill, being used as a source of spares. 25 August 1993.

This view gives a better perspective of that giant nameplate attached to 91017, arriving with the train seen in the previous image. 25 August 1993.

Basking in the summer sunshine is 91022 *Robert Adley*, just after arrival from the north. This loco (as 91122) was scrapped during 2022. 29 July 1994.

Prior to receiving the name *Durham Cathedral*, this is 91002, just arrived at the station. Unfortunately, this loco (as 91102) was scrapped during 2021. 22 July 1992.

Moving down the East Coast Main Line now and forward some years for a series of images taken at Alexandra Palace in the space of one day. This is pioneer 91101 *Flying Scotsman*, now adorned with a fabulous unique nameplate, rapidly approaching the station with a northbound express. This is a modified version of the Virgin livery that was current at the time. The loco is still in service at the time of writing but is wearing an LNER/Inter-City hybrid version of its original Inter-City livery. Those excellent nameplates are still in situ. 13 November 2018.

This is 91113 carrying the then-standard Virgin livery north with another express from King's Cross. This loco was not so lucky, however, being scrapped in 2021. 13 November 2018.

91106 propels a train from the north towards King's Cross. It is displaying decals advertising the Great Exhibition of the North that took place during 2018. This loco is still in front-line service, and now wears the hybrid LNER/Inter-City livery. 13 November 2018.

Another southbound service for King's Cross is propelled by 91103, which is carrying added 'Pride' embellishments. Unfortunately, the loco was scrapped in 2022. 13 November 2018.

Bringing another express south to King's Cross, 91111 *For the Fallen* is wearing another unique livery, this time commemorating the contribution and sacrifices of the armed forces regiments based in the vicinity of the route of the East Coast Main Line during World War One. It was still in front-line service at the end of 2023. 13 November 2018.

Racing north this time is 91112, coming along the straight section of line towards the station as it heads away from the capital. This is one of the most recent examples to be scrapped during early 2023. 13 November 2018.

This is the odd one out amongst the fleet. 91023 was originally released to traffic in late 1990, and later went on to be involved in two very serious crashes first at Hatfield, Hertfordshire, in 2000 and then at Great Heck, North Yorkshire, in 2001. Upon refurbishment in the early 2000s, it should have been renumbered 91123, but a degree of superstition due to its involvement in both these crashes saw it given the number 91132 instead. The loco is heading south towards King's Cross with 'Time to Change' advertisements concerning mental health. However, its own physical health came to an end with scrapping in early 2021. 13 November 2018.

Another express comes up the long straight behind 91127. This is another loco currently in service with LNER in the hybrid LNER/Inter-City livery and is now named *Neville Hill*. 13 November 2018.

The final shot at Alexandra Palace sees 91128 *Intercity 50* heading south for King's Cross. This loco was not so lucky, being scrapped in mid-2022. 13 November 2018.

Moving along the East Coast Main Line still further, we are now at the famous location of Doncaster for the next series of shots, which form the bulk of this section. Firstly, and straight out of the box, we see brand new 91025 racing south on the rear of a King's-Cross-bound service, exactly 28 years before the previous shot was taken at Alexandra Palace. This loco was scrapped (as 91125) during early 2023. 13 November 1990.

During the 2000s and 2010s, the locos went through many changes of operator and livery, and it was very difficult to keep up with them! This is 91120, passing through on the rear of a King's Cross-bound service. The loco is carrying a hybrid East Coast/Virgin livery. In the late 2010s, it received Europhoenix livery and was intended to be selected for export abroad, but this never came about and the loco is currently on a long-term hire to the Crewe Heritage Centre as a static display. 23 September 2015.

In an identical livery, 91122 races through in the opposite direction with a train from King's Cross to Edinburgh/Glasgow. This loco was scrapped in late 2022. 23 September 2015.

There was a period during the early 2020s when LNER was sorting the best examples from the fleet to be used for an extended period of service, mostly on trains to/from Leeds. One of these was 91114 *Durham Cathedral*, seen here being towed out of the works by 08669 *Bob Machin*. The loco is still in service at the end of 2023 but is now carrying the LNER/Inter-City hybrid livery, and still retains its nameplates. 24 February 2022.

Another view of 91114 *Durham Cathedral*; here it has just been shunted by 08724 on to a siding, where it could be statically tested with raised pantographs prior to release back into traffic. It is clear from this image that it has had a full bogie overhaul. 24 February 2022.

08669 *Bob Machin* is seen again, but this time dragging withdrawn and sorry-looking 91104+91108 that are being transferred to another building for further component recovery before being sent for scrap. Although it was withdrawn in February 2020, 91104 is apparently still extant, but 91108 was scrapped in May 2022, not long after this image was taken. 24 February 2022.

This is 91119 *Bounds Green 1977–2017* on the rear of a King's Cross-bound service about to depart. Unlike the other LNER/Inter-City hybrid locos in service today, this one actually carries the near-identical livery it wore from brand new, complete with the *INTERCITY* legend. 12 February 2019.

Another view of the loco showing off to good effect the livery and the excellent cast nameplate. 12 February 2019.

91114 *Durham Cathedral* is seen again, this time awaiting departure with a northbound service from King's Cross. 12 February 2019.

Approaching with another service from King's Cross is 91105 *County Durham*. Both loco and coaches are in the dark blue GNER livery worn from the late 1990s until the mid-2000s. 7 March 2003.

Twelve years later, the same loco is at the same spot, also with a service from King's Cross. The loco was nameless by this time and had received Virgin Trains' red and white livery. It is still in service today, now wearing the hybrid LNER/Inter-City livery but without a name. 23 September 2015.

This is 91120 waiting to depart on the rear of a King's Cross-bound express. The status of this loco was detailed earlier. 12 February 2019.

Double-headed 91s have always been an unusual sight and this one almost caught me out! I just managed to get the camera in position as GNER pair 91104 *Grantham* + 91128 *Peterborough Cathedral* raced through the down centre line with an express from King's Cross. As the pantograph on the train loco was dropped, this spectacle was obviously brought about by a loco failure on this occasion. 4 April 2002.

Quite a common occurrence, however, was the loss of the fairing below the buffer beams, making the locos look somewhat different. This is 91118 *The Fusiliers* racing south on the rear of a King's Cross-bound train. Interestingly, the loco only carried this name for five years from 2018 until being scrapped in 2023. 12 February 2019.

A mix of former East Coast loco and Virgin stock this time as we see 91130 approaching with a train from King's Cross. This was during a 12-year period when the loco was running un-named. This loco is still in service today, now carrying the name *Lord Mayor of Newcastle* and wearing hybrid LNER/Inter-City livery. 23 September 2015.

Right: A nice uniform rake this time with 91128 *Peterborough Cathedral* and stock all in GNER dark blue livery, arriving from King's Cross. This loco was scrapped in mid-2022. 7 March 2003.

Below: The excellent one-off livery applied to 91111 *For the Fallen* can be fully appreciated here as it arrives from King's Cross. 12 February 2019.

This was another one-off; this deep purple livery was applied to 91101 during the mid-2010s. Although not strictly named, the 'Flying Scotsman' legend is boldly emblazoned along the bodywork. In this picture it is arriving from King's Cross. The loco has since received the hybrid LNER/Inter-City livery and is now adorned with exceptional *Flying Scotsman* cast nameplates. See also page 14 (above) earlier in this section. 23 September 2015.

Locos could very often be found stabled by the station in the sidings at Doncaster, and sometimes still can. Just three months old at the time, 91018 already looks rather filthy. It was scrapped (as 91118) in 2022. 13 November 1990.

Eleven years on, this is 91109 *Samaritans*, soon after refurbishment and repaint into GNER colours, being trial run in the sidings. Interestingly, the loco is simply carrying *Samaritans* names rather than *The Samaritans*. Still in service, it now bears the name *Sir Bobby Robson* and is in the hybrid LNER/Inter-City livery. 29 August 2001.

Right: Here we see 91127 propelling a King's Cross-bound service away from the station. This is another loco still in service today, named *Neville Hill* and in hybrid LNER/Inter-City livery. 24 February 2022.

Below: The Virgin livery carried by locos and coaches up to the end of the 2010s looked quite smart and suited the 91s well, especially when clean. 91126 arrives from King's Cross. Unfortunately, this particular loco was scrapped in 2021. 23 September 2015.

Another uniform Virgin rake arrives from King's Cross behind 91103. This loco was one of the first examples to be withdrawn from service in 2019, and was scrapped in the middle part of 2022. 23 September 2015.

Left: On the same day as the previous shot, this is 91104 in East Coast/Virgin livery, also arriving from King's Cross. This loco was withdrawn during 2020 but is believed to be still extant and used for spares. 23 September 2015.

Below: 91124 departs on the rear of a southbound service. This is another one still in service today, and is now in the LNER/Inter-City hybrid livery. 24 February 2022.

Now, a pair of shots showing the changing fortune of 91108, the first withdrawn loco. Firstly, in this image it has just arrived with an express from King's Cross. Withdrawal from service came five months after the date of this shot in July 2019. 12 February 2019.

Right: Just over three years later, the same loco is being dragged out of a works building after undergoing component recovery. Once useable parts had been taken, it was off to the scrapyard from where final destruction took place three months after this shot was taken in May 2022. A sad end. 24 February 2022.

Below: Now for a real bit of déjà vu in the next two shots! This is three-month-old 91019 *Scottish Enterprise* in its original livery propelling a King's Cross service away from the station. The nameplates attached at this time are of the flimsy 'tin' variety. 13 November 1990.

At the same spot, and after countless similar services during the intervening years, 91119 *Bounds Green 1977–2017* propels another express towards King's Cross, incredibly, almost 32 years later! The nameplates affixed now are proper cast ones. Comparing these two shots, it is hard to tell that the loco has seen that amount of front-line service and travelled literally thousands of miles in the interim. 24 February 2022.

Another magnificent one-off repaint adorns the speed record holder 91110 *Battle of Britain Memorial Flight*, seen here departing southbound. Although not the best view due to the glint on the bodyside, take a look at the next shot for a close-up. The excellent cast nameplates are based on the design used on the former Southern Railway Bulleid 'Battle of Britain' class steam locomotives. 12 February 2019.

This is part of the wonderful design on the bodysides of 91110. The opposite side depicts a Lancaster bomber, also part of the famous historic flight that is currently based at RAF Coningsby in Lincolnshire. 12 February 2019.

A smart-looking 91113 *County of North Yorkshire* awaits departure south. This loco was scrapped in 2021. 7 March 2003.

Last built 91131 *County of Northumberland* is just drawing to a halt with a northbound service from King's Cross. This was the last loco to be built at Crewe Locomotive Works and is now preserved as a static exhibit at Bo'ness (West Lothian), Museum of Scottish Railways. 7 March 2003.

91106 glides into the station with a service from King's Cross. Currently still in service, this loco now carries the hybrid LNER/Inter-City livery. 24 February 2022.

Concluding this section on Doncaster, we see a very clean 91107 *Skyfall* on the rear of a King's Cross-bound service. For a short while, this loco carried a special light blue livery to celebrate the release of the James Bond film *Skyfall* in late 2012 and also carried its original number 91007. It is still in service today, retaining its nameplates, but is now painted in the hybrid LNER/Inter-City livery. 23 September 2015.

We now move north of Doncaster to York for our next section, where we see virtually brand new 91001. Unfortunately, I could find no details of this working or indeed a date. However, as the loco is not carrying its later *Swallow* nameplates, it must have been early in its career as it was only nameless between April 1988 and September 1989.

Under the vast roof of this famous station, this is 91005 *Royal Air Force Regiment* in the classic position alongside platform 5 as it awaits departure north with a King's Cross to Edinburgh train. This loco (as 91105) is still in service today, occasionally still visiting York, and is in a very similar LNER/Inter-City livery. 10 April 1996.

Taken on the same day as the previous shot is 91024 *Reverend W Awdry*, just coming to a stop on the rear of a King's Cross-bound service. This loco is also still in service, wearing the same hybrid LNER/Inter-City livery as the loco in the previous shot, but no longer carries a name. 10 April 1996.

Here is 91107 arriving with a terminating service from King's Cross. The loco and coaches are in another hybrid livery of GNER/National Express that became quite common at this time. The fumes above the loco are from a passing freight train and not the 91! The loco is still in service, though now in the hybrid LNER/Inter-City livery and named *Skyfall*. 22 February 2008.

Virgin-liveried 91109 *Sir Bobby Robson* is awaiting departure for Edinburgh with an express from King's Cross. This loco is still in service, once again painted in the LNER/Inter-City livery, but also retaining the nameplates seen here. 23 August 2018.

A hybrid livery again, this time East Coast/Virgin as 91128 slowly draws to a standstill after arriving with a terminating service from King's Cross. At the time of writing in late 2023, this remains the only daylight timetabled Class 91-hauled service north of Doncaster. This loco, however, was scrapped during 2022. 23 September 2015.

Another view under that cavernous roof as 91129 *Queen Elizabeth II* awaits departure north with a train from King's Cross. Once again, the loco and stock carry the crossover livery from GNER to National Express, which basically involved the addition of a white bodyside stripe with the National Express title within it. The vaguely legible name is inscribed in tiny letters in the middle of the loco, also within the white stripe! The loco itself was unfortunately reduced to a pile of scrap metal during 2021. 21 February 2008.

It is amazing how quickly things can change on today's railway system. Although taken not that long ago, this shot is long since history. The two main forms of ECML motive power at the time are seen here, and both of these have since been replaced by Class 800/801/802 'Azuma' units. 91132 departs north for Edinburgh, whilst HST power car 43319 arrives on the rear of a service from Inverness. The 91 was scrapped during 2021 and the power car has been in storage at Ely Papworth Sidings since April of that year.

This is 91107 *Skyfall* again, this time on the rear of a King's Cross service arriving at the main up platform. The loco is still in service, now in the hybrid LNER/Inter-City livery and still carrying its nameplates. 23 August 2018.

A subtle difference with this hybrid livery as 91109 *Sir Bobby Robson* stands on the rear of a King's Cross service. The base colour on this loco is actually silver rather than the more usual grey that was used by the former East Coast. The loco is still in service, now in the hybrid LNER/Inter-City livery and retaining its nameplates. 23 September 2015.

91117 *West Riding Limited* is arriving with the terminating service from King's Cross seen earlier in this section. Although this loco was withdrawn from service in late 2019, it is now used as a source of spares to keep the remaining locos in service. 23 August 2018.

91121 has recently arrived from King's Cross and awaits departure north for Edinburgh. This loco was one of the most recent ones to be scrapped, in early 2023. 23 August 2018.

Judging by the front end, 91112 has had a recent altercation with a bird as it stands ready to leave with a northbound express. This was another loco recently scrapped in early 2023. 23 September 2015.

Right: 91117 *Cancer Research UK* is another loco that carried those ridiculously small 'nameplates' in the middle of the white bodyside stripe. As mentioned previously, this loco is currently being used for spares. 22 February 2008.

Below: In full GNER livery, including the coaching stock, this is 91110 *David Livingstone* calling with a King's Cross to Edinburgh train. This loco has since progressed to become a celebrity, as seen earlier in this book. 7 March 2003.

Taken from a low angle, this shot really emphasises the massive overall roof as 91132 awaits departure north. The former 90023, was scrapped in March 2021. 23 September 2015.

Left: The addition of white buffer rims on 91127 really seems to enhance the appearance of the loco as it arrives with a terminating service from King's Cross. This loco is still in service today, named *Neville Hill* and in the hybrid LNER/Inter-City livery. 25 February 2022.

Below: Pioneer 91101 *City of London* gets away from the station at the rear of a King's Cross-bound train. Those tiny 'nameplates' are (just about) visible in the middle of the loco. The loco is currently in service but wearing a unique livery. 21 February 2008.

91126 *York Minster* is appropriately propelling away from the station with a King's Cross service. Tiny 'name' once again, but there is a prominent full running number stencilled on the front. The loco was scrapped in 2021. 20 February 2008.

Moving a few miles north of York for a selection of images taken whilst having a brief stay at 'The Sidings' hotel, which is well known in the enthusiast's circle for being formed of old railway coaches converted to static accommodation. It is located right next to the ECML at Shipton by Beningbrough in North Yorkshire. Back in this time period, there was an almost constant stream of 91s heading north and south on Anglo-Scottish expresses. All these shots are taken from the viewing area, literally by the lineside, and depict the locos with those tiny 'nameplates' again! This first shot sees 91101 *City of London* racing north with a King's Cross to Edinburgh/Glasgow service. This loco is still currently in service but is in a one-off livery with the name *Flying Scotsman*. 22 February 2008.

We now see 91104 *Grantham,* also heading north with an express from King's Cross. The dark blue livery looked quite good when loco and stock all carried it. This loco is currently being used as a source of spares. 22 February 2008.

Streaking south is 91111 *Terence Cuneo* on the rear of a King's Cross-bound service. See page 25 (below) for how this loco now appears. 22 February 2008.

Another northbound service zooms past in the capable hands of 91129 *Queen Elizabeth II*. Unfortunately, this loco was scrapped during 2021. 22 February 2008.

The last shot here sees 91122 *Tam the Gun* going south on the rear of a King's Cross-bound service from Edinburgh/Glasgow. The loco was scrapped during 2022. 20 February 2008.

Yet to be refurbished, this is 91017 *Cancer Research UK* standing at Edinburgh Waverley after arrival from King's Cross. Many examples, if not all, received the GNER livery before they were refurbished. This is another loco being used for spare parts at the time of writing. 29 February 2000.

Also at Edinburgh Waverley, this is 91027 just arrived from King's Cross. The loco was un-named at this time and although it doesn't look it, was just 18 months old. It is currently still in service carrying a hybrid LNER/Inter-City livery very similar to how it appears here, plus the name *Neville Hill*. 24 August 1992.

To end this section on Class 91s, we are back where we started at King's Cross, where we see 91027 some years later, after it had gained GNER livery but with the very short-lived white lettering. It had also recently lost its *Great North Run* nameplates. 3 April 1997.

On the same day as the previous shot, this is original speed record holder 91031 in a fresh coat of dark blue paint, also displaying white lettering. It had also just lost its *Sir Henry Royce* 'tin' nameplates. The loco is now preserved as a static exhibit. 3 April 1997.

Another couple of years on and gold lettering had become the norm. Here we see 91021 *Archbishop Thomas Cranmer* awaiting departure with an express for the north of England. A few years after, around 2011, when GNER took over the running of these locos, the naming policy came back into favour, albeit with these new-style sticker efforts. This loco was scrapped in early 2023. 27 February 2000.

Before becoming a celebrity later in its career, 91119 *County of Tyne & Wear* was just another loco in the fleet. It has just arrived with a train from Edinburgh/Glasgow. 7 March 2003.

More line-ups now to finish off. This is 91122+91111+91125 all in National Express/GNER livery, having arrived with trains from the north. 20 February 2008.

Whilst turning the camera slightly, we see 91122+91127+91101 almost making a full house. Pity I didn't have a very wide-angle lens at the time! 20 February 2008.

Both locos and stock are seen here in what was a rather pleasing light grey, set off by a purple stripe. This was the East Coast livery, almost the opposite of the GNER dark blue, and stood out quite well. 91121 has just arrived, whilst 91124 gets under way with a Leeds service. 23 July 2014.

A few years before, we see 91131 in standard light grey. Although also in East Coast colours, 91109 next to it is more silver than grey. 18 August 2012.

Virgin colours this time as we see 91108 just arrived and 91127 just departing. 23 August 2018.

This section on 91s comes to an end with this shot of 91009 *Saint Nicholas* next to 89001. At this point, the sole Class 89 was about a week into its hire period with GNER for use on services to Leeds, whilst the 91 must have been in its final few days in Inter-City livery before also receiving GNER colours. 3 April 1997.

Chapter 2

Class 92

The Class 92 is a powerful dual-voltage electric locomotive that can run on 25kV using overhead wires or from 750V in third rail-electrified areas. Outwardly, these are very similar to the Class 60 diesels but are designed primarily to operate through the Channel Tunnel between the UK and France. The fleet eventually totalled 46 locos but, unfortunately, the volume of freight envisaged for them to work through the Channel Tunnel never really materialised, nor did the accompanying 'Nightstar' sleeper trains even start. At one time, it was hoped that they would be working freights from France to places such as Eastleigh, but once again that didn't transpire. More recently, however, a few have been employed on sleeper trains to/from London Euston to Scotland, on which they seem reasonably reliable. Many members of the class have also been exported to work abroad, whilst others have been stored for many years, unlikely to work again. For all of this latter-day upheaval, the class finds itself still intact and to date no examples have been scrapped.

To start this section, we see 92005 *Mozart* **heading north through Watford Junction with an unidentified service. This loco has since been exported to Croatia and renamed** *Emil Cioran.* **4 July 2002.**

Looking in the other direction at Watford Junction, this is 92022 *Charles Dickens*, heading for Wembley with a southbound Intermodal service. This was one of the few locos that actually carried cast nameplates. It has since been exported to Bulgaria. 4 July 2002.

Also approaching Watford Junction is 92034 *Kipling* with a train of flat wagons, probably also bound for Wembley Yard. This loco has since been exported to Bulgaria, where it is in everyday traffic. 26 July 2006.

Moving to Rugby, we see 92039 *Johann Strauss* waiting to head north with another unidentified train. This loco was also exported, but this time to Romania. 12 April 2002.

A whole train of liner flats is always a difficult subject to photograph. This is 92015 *D. H. Lawrence* coming to a stand at Rugby with a poorly loaded train on what was then the up avoiding line. The station here has since been rebuilt and where the train is has now become a whole new island platform. This loco is one of a handful to have received the DBC red livery and is still currently in UK service. 21 September 2006.

This is 92043 *Debussy*, hauling Freightliner's 90047 southbound through Rugeley Trent Valley station, a slightly unusual sight as the 92 by this time was working for GBRF. It has since been repainted into standard GBRF livery and now carries the name *50 Years Service – Andy Withers*. The Class 90 is still in service with Freightliner, carrying that company's latest orange and black livery. 4 February 2016.

Plain grey 92002 *H. G. Wells* is seen at Carlisle, about to head south with an Intermodal service. This loco has since been exported to Croatia, where it now carries the name *Lucian Blaga*. 2 February 2004.

Stafford now, with the unusual sight of a pair of 92s. This is 92041 *Vaughan Williams* + 92034 *Kipling* slowly approaching the signal in the up through line with a northbound Intermodal train. With the lowered pantograph on the train loco, this looks like a failure has occurred. Also note that both locos had received the EWS 'big beastie' decals on the bodysides. 23 January 2009.

Another shot of the pair seen in the previous shot, clearly showing the lowered pantograph. 92034 is now at work in Bulgaria, whilst at the end of 2023, 92041 has recently had a test run, with reinstatement to traffic to follow with DBC. 23 January 2009.

An unusual sequence of shots now as firstly we see EWS 'big beastie' pair 92009 *Elgar* + 92015 *D. H. Lawrence* stabled in the little-used up bay platform, unusual in itself. 23 January 2009.

Later in the day I discovered the reason for the locos' presence here when they headed for the sidings just east of the station and, after a bit of manoeuvring, came back with a crippled wagon sandwiched between them. Careful study will reveal a wheel skate under the first set of wheels behind 92015. 23 January 2009.

Here we see 92009 now leading this strange ensemble and about to depart for the short run up to Crewe, probably restricted to about 15mph or so. Both of these locos are still in the UK with DBC, although 92009 is currently stored. 23 January 2009.

On the same day as the last sequence of shots, this is 92036 *Bertolt Brecht* passing through with a southbound Intermodal. This loco is still currently in service with DBC in the UK. 23 January 2009.

Occasionally when a loco has been worked on at Crewe, it will be given a test run to Stafford to check all is OK. Looking smart in DB livery, 92042 arrives on such a run. 4 February 2016.

Upon arrival, the 92 performed a few static tests, raising and lowering the pantograph and then returning to Crewe. The impressive size of these locos can be seen here. This particular one is still at work in the UK with DB. 4 February 2016.

The very first example to receive DB red livery was 92009, which is seen here in absolutely immaculate condition passing through Kensington Olympia on what I think was its first train in this livery, the 6B20 Dollands Moor to Wembley. 27 May 2011.

Later the same morning, I managed to get this further image at South Kenton of its follow-on service, 6B41, the 11.43 Wembley to Daventry. Upon repainting, the loco lost its *Elgar* nameplates, but a short while later it received the cast nameplates *Marco Polo*. It has, however, been out of service for some time now. 27 May 2011.

A somewhat grabbed shot, as it caught me out. 92008 *Jules Verne* is just about captured hurrying south through Tring with a mixed freight. Although still on DB books, this loco was put into storage as long ago as December 2006 and is highly unlikely to work again. 10 July 2003.

This is 92011 *Handel* stabled between duties at Rugby. The plain two-tone livery did very little to enhance these locos. This example, however, is still in service with DB and, although still in plain two-tone grey, it does still sport an obsolete EWS 'big beastie' logo! 23 July 2004.

Back to Kensington Olympia now, where we see 92041 *Vaughan Williams* passing through with a Dollands Moor to Wembley container train. This is another loco still in service with DB at the time of writing. *Circa* late 1990s.

Unfortunately I have mislaid my notes regarding this somewhat less than sharp image, but is included due to being very unusual. 92008 *Jules Verne* is hauling three 'dead' Class 66s in the consist of a Wembley-bound Intermodal train passing through Kensington Olympia. Back in those days, it was rare to see this many locos on the front of a train. Circa late 1990s.

One way to try and make a long train of empty flat wagons look good is to compact the perspective with a zoom lens. This is 92003 *Beethoven* coming slowly through Kensington Olympia with 4E32, the 11.52 Dollands Moor to Scunthorpe empty steel train. This loco is currently in service in Romania. 27 July 2010.

Just over a year earlier, 4E32 is seen at the same location, this time behind 92007 *Schubert*. This loco has been in storage since August 2011. 29 October 2009.

Most southbound freight trains at Kensington Olympia take the platform line, but this is the unusual sight of a southbound service taking the through line. 92016 *Brahms* has a rake of covered Cargowagons heading from Wembley to Dollands Moor. This loco has been stored since September 2017. 10 July 2003.

During the late 2000s, Class 350 electric units were being delivered to the UK and were hauled up from Dollands Moor, usually by a diesel loco. However, 92s were also occasionally used for these drags, to which end 92015 *D. H. Lawrence* is hauling brand new 350235 through, plus many other wagons. Both the loco and EMU are still in service today. 13 November 2008.

This time we see 92022 *Charles Dickens* passing through with the 6B20 Dollands Moor to Daventry service. This loco is now in service in Bulgaria. 27 July 2010.

92017 *Shakespeare* is waiting to depart Kensington Olympia light engine to Dollands Moor. This loco went on to be a minor celebrity later in its career (see page 67 (above)) but was stored in March 2012. 10 July 2003.

A few shots now of open days where 92s have often made an appearance. The pioneer loco 92001 *Victor Hugo* was one of only two examples (along with 92031) to receive EWS livery. It is seen here on display at Crewe Works open day. This loco now operates in Romania and has been re-named *Mircea Eliade*. 31 May 2003.

Here we see two-tone grey 92035 *Mendelssohn* at the same open day as the previous image. Incredibly, this loco has now been stored for almost 20 years and is unlikely to run again. 31 May 2003.

An earlier open day at Crewe sees 92005 *Mozart* in the works building under maintenance. This loco has since been exported to Croatia and renamed *Emil Cioran*. 20 May 2000.

In the same building is 92018 *Stendhal*, complete with its SNCF insignia below the cab windows. This was originally one of the locos owned by French railway company SNCF in the early days. Nowadays owned by GBRF, it is in deep blue livery and is used on the Anglo-Scottish Caledonian Sleeper overnight services. 20 May 2000.

Our last shot at this particular Crewe open day is of another plain two-tone grey loco, 92026 *Britten*. This loco is now at work in Croatia. 20 May 2000.

Nowadays, the use of 92s in third rail-electrified areas is not that common, but on this day I photographed two trains on the same morning within a short time of each other. The location here is just east of Hollingbourne in Kent as we see GBRF-operated 92032 *I Mech E Railway Division* approaching with a rake of brand-new oil tanks as 6M92, the 10.21 Calais Frethun to Wembley (loco change) and then on to Lindsey. This loco is still in service with GBRF. 23 August 2012.

Not long after the last shot, we see 92015 at the same location with the regular 4E32, the 11.52 Dollands Moor to Scunthorpe empty steel train. This loco is still in service with DB today. 23 August 2012.

This is 92042 *Honegger* passing Sevington loop, just east of Ashford International, with 4E32, the 11.52 Dollands Moor to Scunthorpe empty steel train again. HS1 can be seen behind the train used by Eurostar trains and Class 395 Southern trains. As noted earlier in this book, 92042 has since been painted in DB red livery. 8 October 2010.

4E32 features strongly in this section, as it was the last regular Class 92 daylight diagram to run up from Dollands Moor, and was at a good time of day! 92009 *Elgar* is passing through Ashford International. This is another loco to gain the DB red livery and has since been renamed *Marco Polo*. It is, however, out of service at the time of writing. 12 November 2009.

Here we see 92001 *Victor Hugo* passing through the London suburbs at Denmark Hill, again with 4E32. This loco has since been exported to Romania and is now named *Mircea Eliade.* 19 May 2011.

Looking for all the world like a light engine working, more careful study actually reveals that 92026 *Britten* is once again hauling 4E32, this time coming across Eynsford Viaduct between Tonbridge and Swanley. This loco is now at work in Croatia. 18 November 2010.

Another view of Eynsford Viaduct, this time from the public foot crossing that goes over the line at the western end of the structure. This is good old 4E32 again, this time being hauled by 92034 *Kipling*. This loco is now at work in Bulgaria. 29 September 2011.

During 2009, 92017 (formerly *Shakespeare*) was repainted into Stobart livery and renamed *Bart the Engine*. It is seen here approaching Wandsworth Road on a miserable damp autumn day. This loco has been stored since March 2012. 11 November 2011.

92042 is seen again with 4E32, passing Wandsworth Road station. This is a popular location for enthusiasts, but the main view is looking in the opposite direction towards London Victoria. 17 October 2012.

Between Clapham Junction and Kensington Olympia is Cremorne Bridge, which carries the line over the river Thames near Chelsea Harbour. This time, 92012 *Thomas Hardy* is captured as it glides across with 4E32 heading north. This loco is now at work in Hungary and has been re-named *Mihai Eminescu*. 5 November 2009.

Our penultimate look at 4E32 sees the first of the class again, as 92001 *Victor Hugo* threads its way through Bickley station on its way north. Quite a few of the stations in this area have the station buildings built across the tracks in the manner seen here. 9 December 2009.

A regular at the time on 4E32 was 92026 *Britten*. It is seen here approaching the station at Shortlands with a good rake of flats in tow. 29 July 2010.

The first of a few views of a popular (albeit temporary) service. Taken from the footbridge seen in the background of the previous shot, this is Europorte-branded 92038 *Voltaire* running around 30 minutes before the previous shot with 6Z93, the Dollands Moor to Wembley crew training run for GBRF. This was a regular working for a while at this time and was sometimes top-and-tailed by Class 92s. This loco is still operational with GBRF although now carries Caledonian Sleeper dark blue livery. 29 July 2010.

Moving further east, we catch up with 92044 *Couperin* passing through the up through line at Paddock Wood with 6Z93. This loco is still in service with GBRF. 25 March 2010.

This is the scene at Tonbridge as 92043 *Debussy* gets away from a signal check with 6Z93, which will be routed via Sevenoaks from here. This loco is still in service with GBRF but now wears the standard GBRF livery and carries the name *50 Years Service – Andy Withers*. 17 March 2010.

The following day, 92043 *Debussy* is seen again, this time approaching Shepherd's Bush with the returning 6Z93 Wembley to Dollands Moor. The train kept the same headcode heading back east and was in effect the same train. 18 March 2010.

Back at Kensington Olympia again, we see 92038 *Voltaire* top-and-tailing 92043 *Debussy* on their way past with 6Z93. The top-and-tailing mode was really just to avoid running a loco round at Wembley. 27 July 2010.

Another view as the train heads away from the camera, with 93043 *Debussy* bringing up the rear. 27 July 2010.

To conclude this section on Class 92s, we have a couple of shots on the ECML, along which they have never been common. This is 92009 *Elgar*, stabled at Doncaster. The loco had just worked Hertfordshire Rail Tours' 'The Wig & Weasel' charter from Finsbury Park but, upon arrival at Doncaster RMT (for a booked loco change), it was declared a failure. The problem must have been resolved quite soon, as it must have run the short distance to the up bay, as seen here. I think this may well have been one of the first times a 92 had worked a charter on the ECML. 8 March 2003.

Still at Doncaster, GBRF Caledonian Sleeper-liveried 92014+73969 are passing through with 0Zxx from Craigentinny to Doncaster RMT. This pair were heading south for maintenance and were eventually hauled by 47749 *City of Truro* across to Leicester LIP. Having lived all my life in the south of England, it is still rather strange to see Class 73s working in Scotland! 24 February 2022.

Chapter 3
Mk4 DVTs

These Driving Van Trailers (DVTs) were designed specifically to operate with the Mk4 coaching stock that operated on the ECML, worked by Class 91 locos. They were mostly located on the southern end of these sets so they were on the rear heading north. A few are still in use with LNER, while others have moved to Transport for Wales (TfW) for Holyhead to Cardiff trains, using a short rake of Mk4 coaches hauled/propelled by a Class 67 diesel.

When focusing on loco-hauled trains, many times the Driving Van Trailer (DVT) that is used on the rear of the coaching stock is often ignored. Based on the design of the Class 91, the Mk4 DVTs are exclusively used with the Mk4 coaching stock. 82212 and 82219 are seen 'on the stops' at London King's Cross on the rear of northbound departures. 82212 is still in service with LNER at present, whilst 82219 was scrapped in 2021. 23 August 2018.

This is 82226, heading north past Alexandra Palace on the rear of an express from King's Cross. Since becoming redundant from ECML services, this example has transferred to Transport for Wales (TfW) and is used on Holyhead to Cardiff services powered by hired-in Class 67 locos. 13 November 2018.

Another shot at Alexandra Palace station as 82217 (formerly *Off to The Races*) speeds south at the front of a King's Cross-bound express. This example was de-named some years ago and scrapped in mid-2020. 13 November 2018.

Here we see 82205 (formerly *The Flying Scotswoman*) in its special Flying Scotsman livery, again heading south at Alexandra Palace on its way to the capital. It was originally intended that this DVT would always work with similarly liveried 91101 but this didn't always happen! It is still in service with LNER at present. 13 November 2018.

82211 is seen at the same location with another King's Cross-bound service. This is another one still in service with LNER. 13 November 2018.

This is pioneer 82200 on the rear of a Leeds-bound service calling at Doncaster, which gets a series of shots for this station under way. After being removed from ECML services, this was one of a few examples of the class that transferred to Grand Central, which was planning to run services from London Euston to Glasgow that ultimately never materialised. However, this DVT did receive Grand Central black livery, but instead moved to TfW. 23 September 2015.

The same DVT is seen here having been dragged from a building at Doncaster Works. This was during the short while that it had no further use before the above events. 12 February 2019.

This is a Leeds to King's Cross service weaving its way into Doncaster station behind 82229. This DVT was taken on by TfW and is still in service on the Holyhead to Cardiff route, now powered by Class 67s. 12 February 2019.

82218 arrives on the rear of another King's Cross to Leeds service at Doncaster, on a fine autumnal morning. This example is currently in storage at Worksop, with an uncertain future. 23 September 2015.

A King's Cross-bound service departs south headed by 82201, looking rather work-worn. This DVT is currently with TfW but, at the time of writing, is in storage. 23 September 2015.

82205, in the rather pleasing purple livery it carried during the mid-2010s, is arriving on the rear of a King's Cross to Leeds service. It is still in service but now wears a modified LNER/Flying Scotsman livery. 23 September 2015.

And here it is in the same spot four years later, also on the rear of a King's Cross to Leeds train. Incidentally, the haze across the station was emanating from a nearby fire away from the railway; nothing serious, thank goodness! 12 February 2019.

82228 is bringing up the rear of a non-stop express bound for Edinburgh/Glasgow as it races north through the station. This is one of the examples since scrapped, in this case during 2020.
23 September 2015.

82217 is captured here as it leads a King's Cross train away from the station. This was also scrapped during 2020.
12 February 2019.

Rewinding a few years to GNER days, we see 82217 again at almost the same spot, also leaving with a King's Cross-bound train.
7 March 2003.

Arriving at the adjacent platform is 82218, again heading for the capital. The usual smattering of enthusiasts can be seen on the main down platform. 7 March 2003.

On a chilly New Year's Day, 82219 is on the rear of a Leeds-bound service calling at the station. The effect of the whole train being formed of uniform GNER stock can be appreciated. This one was scrapped during 2021. 1 January 2004.

This time, we see 82210 approaching with a train from Edinburgh/Glasgow. This DVT is currently stored, probably as a source of spares for LNER. 12 February 2019.

82223 awaits departure south for King's Cross whilst 82203 arrives on the rear of a King's Cross to Leeds train. Of this pair, 82223 is still in service with LNER but 82203 was not so lucky, succumbing to the scrapman during 2020. 23 September 2015.

At the north end of the station again, we see 82209 approaching with a King's Cross-bound service. Unfortunately, this is another example that has been scrapped, being cut up in 2022. 12 February 2019.

Arriving from King's Cross is 82230 on the rear of another Leeds service. This DVT is now with TfW but at the present time is in storage. 23 September 2015.

82220 departs south for the capital. Many of today's trains are seriously lacking in luggage space, but these DVTs had plenty of it. This example is on the books of LNER but it is presently in storage. 12 February 2019.

Racing through the up fast line is the last of the type to be built, 82231, with a Glasgow/Edinburgh to King's Cross service. As of late December 2023, this DVT is in storage by LNER, located at Worksop. 23 September 2019.

82202 is in the hybrid East Coast/Virgin livery it carried for a while, awaiting departure south for King's Cross. This one was scrapped during 2021. 23 September 2015.

Standing at the same spot some seven years later is 82211, leading another southbound service. The white-edged buffers are a nice touch. It is still in service with LNER. 24 February 2022.

A full set of loco/coaches/DVT in Virgin livery was quite striking, as can be seen here. 82208 is on the rear of a Leeds service just arrived from London. 12 February 2019.

Turning the clock back a few years, we see 82215 and a full GNER rake hurtling up the through line with an Edinburgh to King's Cross express. This DVT is still currently in service with LNER. 7 March 2003.

Looking north again, we see 82211 rapidly approaching with an Anglo-Scottish express. It is overtaking another type of train that has since disappeared from the national network, the humble Pacer, in the form of 144006. 12 February 2019.

At the same spot, we see 82226 departing on the rear of a northbound service, whilst speeding past it is 82212 on its way to King's Cross. 82226 is now with TfW but currently stored, but 82212 is still in service with LNER. 12 February 2019.

A nice clean 82213 pulls away with a service for King's Cross. The white buffer rims will again be noted. This example is also still in front-line service with LNER. 24 February 2022.

82219 is about to make the station call with a King's Cross-bound service. Not many of these Mk4 DVTs actually carried names but, for a while during the early 2000s, this one was named *Duke of Edinburgh* (sticker type) when operating for GNER. Unfortunately, it was scrapped in 2021. 12 February 2019.

Another white-rimmed buffer example, 82222, arrives at the rear of a King's Cross to Leeds train. This one is still in service with LNER. 24 February 2022.

Our final shot at Doncaster sees 82223 arriving with a King's Cross service, whilst to the left is the new order on the ECML in the form of 800113. All Anglo-Scottish expresses are now in the hands of these 'Azuma' units. 12 February 2019.

We have now reached York for another series of shots. This is hybrid East Coast/Virgin-liveried 82208 departing on the rear of a King's Cross to Edinburgh train. Although this one is still in use with LNER, it only operates between London and Leeds, or occasionally to York. 23 September 2015.

Left: GNER-liveried 82202 is awaiting departure on the rear of another northbound service. This example was not so lucky, being scrapped during 2021. 22 February 2008.

Below: This is 82230 departing southbound for King's Cross. Currently this DVT is in Grand Central livery but with TfW in storage. 20 February 2008.

Above: GNER-liveried 82211 is captured coming around the sweeping curve into the station from the north with a southbound express. 22 February 2008.

Right: Ten and a half years later, the same DVT is now in Virgin livery and is at the same position with another Anglo-Scottish express. It is still in service with LNER, though is much less frequently seen at York. 23 August 2018.

82213 is another one seen in the same spot as it arrives from the north. This example is also still in service with LNER. 22 February 2008.

A day earlier, 82213 was working a similar duty as it arrives at the platform. By this date, most of these DVTs had done thousands of miles on ECML service. 21 February 2008.

Seven years on and thousands more miles later, 82213 is now in Virgin livery and departing on the rear of a Glasgow/Edinburgh-bound train. 23 September 2015.

On the same day, 82220 is also heading north on the rear of a train from King's Cross. Although still with LNER, this DVT is currently stored. 23 September 2015.

Part of the wonderful overall roof can be appreciated here as 82219 (formerly *Duke of Edinburgh*) brings up the rear of a train departing for Glasgow/Edinburgh. This one succumbed to the scrap metal merchant in 2021. 23 September 2015.

The last shot at York sees 82226 departing on the rear of another express heading for Scotland. This DVT is currently in operation with TfW and is now wrapped in a vinyl promoting Alzheimer's Society Cymru. 23 August 2018.

Now come three shots taken just north of York during my stay at The Sidings Hotel at the same time as those taken earlier in this volume of Class 91s. Heading south at full line speed is 82205, with a train from King's Cross. This one is still in service, but in the hybrid LNER/Inter-City livery. 22 February 2008.

This time 82227 is in the same spot, also heading south for York and King's Cross. Now in black Grand Central livery, this example is stored by TfW. 22 February 2008.

Last but not least is 82222 with a southbound express. This is another one now carrying the hybrid LNER/Inter-City livery and still working main-line services. 21 February 2008.

To conclude this volume, we will take a quick look at a few DVTs now in operation with TfW. With Hope House Children's Hospice wraps on the bodyside, 82216 is arriving at Newport with 1V91, the 05.30 Holyhead to Cardiff Central service. 67015 was providing power on the rear. 23 November 2021.

82216 is now on the rear of the returning 1W93 11.25 Cardiff Central to Holyhead service, also departing Newport. 23 November 2021.

Carrying a vinyl wrap of its own, but advertising the Lifeboats/Coastguard services, 82229 has just arrived at Cardiff Central with 1V91, the 05.30 Holyhead to Cardiff Central. The stock is still in de-branded Virgin livery from ECML days. 67020 was on the rear this time. 14 March 2023.

The final shot in this volume shows 82200 with its Remembrance Day wrap awaiting departure from Cardiff Central with 1W93, the 11.25 Cardiff Central to Holyhead service. Note the all-over black livery adopted by TfW for these trains. 14 March 2023.

Other books you might like:

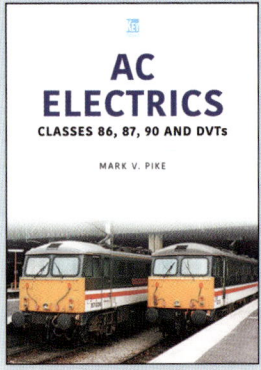

Britain's Railways Series, Vol. 59

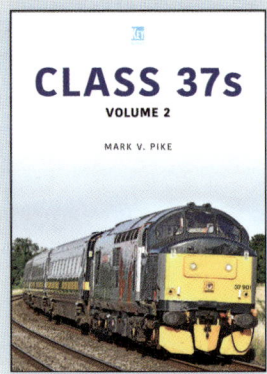

Britain's Railways Series, Vol. 55

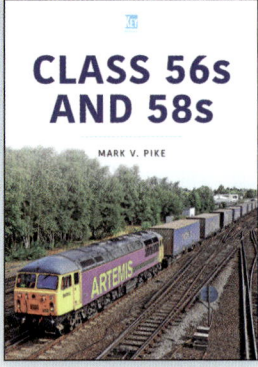

Britain's Railways Series, Vol. 54

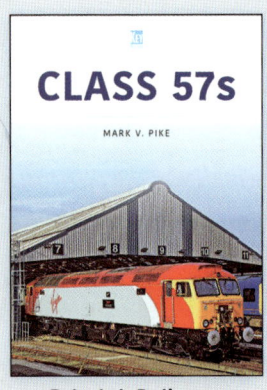

Britain's Railways Series, Vol. 53

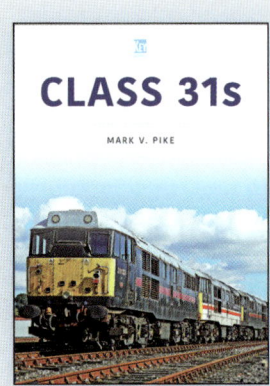

Britain's Railways Series, Vol. 50

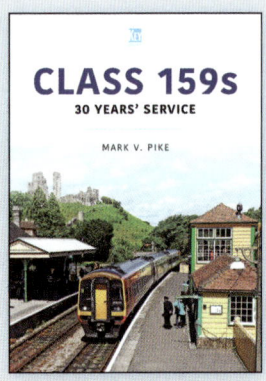

Britain's Railways Series, Vol. 47

For our full range of titles please visit:
shop.keypublishing.com/books

VIP Book Club

Sign up today and receive
TWO FREE E-BOOKS

Be the first to find out about our forthcoming book releases and receive exclusive offers.

Register now at **keypublishing.com/vip-book-club**

Our VIP Book Club is a 100% spam-free zone, and we will never share your email with anyone else. You can read our full privacy policy at: privacy.keypublishing.com